A BOOK ON WATER

In Science and in Religion

V. V. Raman

Create Space

Copyright 2019 V. V. Raman

From the Indigenous Declaration on Water, (July 8th, 2001 in the Musqueam Territory):

We raise our voices in solidarity to speak for the protection of Water. The Creator placed us on this earth, each in our own sacred and traditional lands, to care for all of creation ... to ensure the protection and purity of Water. We stand united to follow and implement our knowledge, laws and self-determination to preserve Water, to preserve life. Our message is clear: Protect Water Now!

TABLE OF CONTENTS
(Y)

PREAMBLE 5
1 WATER AS A SUBSTANCW 9
2 SURGING WAYER: OCEANS 17
3 FLOATING WATER: CLOUDS 25

PREAMBLE

Water was an interesting theme on which one can reflect from the perspective of science and of religion both of which have shaped civilization as water itself has.

One doesn't have to be a student of world's religions to know that there is not a single religious tradition where water hasn't played a significant role. Sprinkling and bathing, sipping and gulping water: all these have entered religious rites and rituals one way or another. There is a charm in simple rituals that defies logical explanation, but rituals add an aesthetic dimension to any proceeding. Even religious naturalism brings in water. The index of a charming book by a religious naturalist on the *Sacred Depths of Nature* has thirteen entries under water. The more one delves into the prayers and practices of religions, the more it becomes clear that though they differ in concrete expressions, though they are varied in their particular cultural manifestations, ultimately there are many things in common among the religions of the human family.

Whether Judaism or Christianity, Islam or Hinduism, Buddhism or Native American religions, water comes in one way or another. Even Unitarians, who are not normally thrilled by rituals, sometimes begin their fall season in a service to which the congregation brings waters from the places they visited during the summer. Over the years I have dipped my hands in the sacred Ganga of India, in the Trevi Fountain in Rome, in the the Arabian Sea, in the Mediterranean, in the Indian, Atlantic, and the Pacific Oceans, in the Caribbean, in the Seine in Paris and Avon at Stratford, in the Douro River in Porto, the Tennessee River, in the Mississippi in Omaha Cuyahoga River in Akron, and the Snake River in Iowa.

 I reflect in these pages on the 00aspects of water in different forms. Science has accumulated lots of information about wate,: from its constitution to its appearance in various states and forms; and their impact on human beings and on our cultures. I also consider how water occurs in some religions.

1
WATER AS SUBSTANCE

Many modes of water

We can picture water in its many modes. Water may stand still as in a lake or in a pond, calm and serene, except perhaps for some ripples caused by breeze or a stone that was made to skip on its surface. Or it may flow gently as in a stream or in a brook in the wilderness, gurgling as it skips over pebbles and other obstacles along the way, inspiring poets to versify. Remember Tennyson's lines:

    I chatter, chatter, as I flow
    To join the brimful river,
    For men may come and men may go,
    But I'll go on for ever.

Or again, water could be gushing in a river, wide or narrow, punctuated by bridges here and there, meandering for miles through little towns and big.

Pascal described rivers as *les chemins qui marchent*: roads that walk. Sometimes the smoothly flowing river becomes a rapid, and then, unexpectedly its bed terminates at some place, it takes a precipitous plunge. The poet Southey described the cataract of Ladore thus:
>The cataract strong, then plunges along
>Striking and raging, As if a war waging
>Rising and leaping, sinking and creeping,
>Swelling and sweeping,
>Showering and springing, and so on.

When this is grand, it attracts tourists and photographers to marvel at the picturesque plunge of a large body of water which suffers the consequence of a terrible surprise, as it were. When one stands in front of the Niagara one cannot but be impressed by the enormous and never-ending onrush of some 750 thousand gallons of water each second from a height of some 170 feet over a brink that stretches to more than 36 hundred feet! It staggers our imagination that this has been going on for thousands of years.

Water can also be roaring and turbulent in the waves of the seas, tossing fragile boats and rafts, sometimes even ocean liners.

We can see all this plainly with our eyes. But water is also continuously rising invisibly as vapor from the surfaces of rivers, lakes and oceans all over the world. And this vaporous water periodically falls back to ground. This may be gentle as a drizzle or strong as the monsoon rain in the tropics. Water may trickle down in meager proportions through a crevice like wealth from the very rich to the very poor. Or, it may flow smoothly down as from a jug into a glass. All these show the versatile aspects of the wondrous thing we call water.

The ancients pictured a watery flow even in the heavens, which is why the whitish patch up on high came to be called *Via Galactica*: Milky Way. This patch struck observers in India as a sacred river in the heavens, which is what is meant by its Sanskrit name *Ákásh Ganga*: the ethereal Ganges. This is a beautiful mythopoetic way of conveying the idea that if water is the source of all life, then heaven is the source of water itself. This vision is not unlike some modern theories by which water on earth had its origins in our planet's periodic encounters millions of times with stray comets which are made up essentially dirt-ridden chunks of ice. Who can tell what unseen forces, what invisible principles gave rise to this marvel we call the physical universe in which we have found our niche for a brief span of time in a slice of endless eternity?

We often think of water only in its liquid form, cool and fresh on a hot summer day, or boiling in a whistling kettle ready to be transformed into a cup of tea or a bowl of soup. But the same flowing water can also be in frozen stiffness in little cubes of ice and on mammoth icebergs in the colder recesses of the planet. There is water in cushiony clouds, there is also water in the air, and there is water deep in the underground too.

On the one hand, water is abundant on earth. After all, there is immensely more land than sea on the surface on the earth. Yet we all know that much of this is sea-water, that is to say, too rich in salts to taste any good. It is as if we have billions of dollars in non-negotiable currency. Good to look at, but useless in the shopping mall.

Recall how Coleridge expressed pithily this awkward human plight when he wrote:

Water, water, everywhere; and all the world did shrink

Water, water, everywhere; nor any drop to drink.

One theme in the modern world is:

> Water, water everywhere, not enough for all to drink,
> This is something of which we all must think.

In our own times, as a consequence of umpteen causes, and most of all, as a result of a horrendous growth in human population, there isn't enough water everywhere, and not enough for all to drink.

This is a problem that is getting to be more and more serious in many parts of the world. The subject is vast as the oceans also.

It is common knowledge that though we imagine our physical body to be made up of flesh and bones, muscles and solid matter, chemically speaking, the human body is 70% just water. This is of course a little less that in the shark whose body is 80% water, or in the lobster which contains 90% water.

It is our intrinsic connection with water that makes thirst such an important aspect of the human condition. Thirst is natural. It is also acute when we are deprived of water for long. Most of all, thirst is never fully satisfied in that it is recurring state for the body. It is the reminder that we are still alive. That is why the Buddha referred to insatiable desires and addictive cravings as thirst, for they too recur even after one is fully satisfied. Human desire can never be *quenched* fully. Then too, we thirst not only for water, but also for love and companionship, for knowledge, and some even talk about the thirst for transcendence. As Ben Johnson wrote,

> The thirst that from the soul does rise
> Doth ask a drink divine.

Water molecules

It is common knowledge the water molecule consists of two atoms of hydrogen and one of oxygen: H_2O. Each hydrogen atom is *covalently*[2] bonded to the oxygen atom via a shared pair of electrons. Quantum chemistry has shown that the angular separation between the bonds plays an important role in many of the properties of water. The electric charges in the water molecule are unevenly distributed. This makes water a polar molecule. It is the polarity that makes solutions in water possible. Bluntly put, no polar molecules, no soup or syrup, because neither salt nor sugar would dissolve in the liquid. More seriously, if there can be no salt solution, we would be missing ourselves, because there would no cell formation, no life at all, here or anywhere. The ultimate existential question which all religions raise is rooted in the fact that the water molecule is polar.

Our chemistry books call water a colorless, odorless, tasteless liquid. But there are many manifestations and subtle properties of water which are enormously valuable, even indispensable for life. Perhaps it is the drab neutrality of water, its apparent quality-less-ness that makes water so essential and versatile. To those who do not inquire into a sage's wisdom, nor reflect on his words, his unostentatious demeanor may give the impression of a calm and colorless individual. So it is with water. Water is normally smooth-flowing and indifferent to those that tamper with its peace, but it is rich in properties that sustain all of life, and does much more.

Water by itself cannot produce life. The salts by themselves cannot become life either. It is somewhat like the physical world by itself cannot produce science. Nor can there be any science if there was only pure consciousness and no physical world. Thus, in a peculiar way water and salts are like religion and science, where by religion I mean all the grand things that emerge from a conscious and feeling entity. Parallel to Einstein's famous quote[4] we might say: Science without religion is like the salt without the water, and religion without science is like water without the salt.

2
SURGING WATER: OCEANS

Introduction: Waves

Every time I have stood firmly on a wet sandy beach, facing the vast blue waters of the sea, my experience has always been the same. Majestic waves always splash incessantly in complete indifference to my presence, lashing on the long shoreline, as if they were part of a furious army charging at a defiant enemy. Then they die away into foamy puddles that meekly recede back into the tumultuous expanse of the ocean. From the safe distance where I usually stand, my legs become wet as far as my knees when the cool waters climb over my frame, but I feel as if the waves are trying to uproot me from my vertical posture. I remember how at one time, a hearty rolling wave caught me unawares. Within a brief moment it overpowered me, threw me down, and tried to draw me towards the vast bosom of the sea. I crawled back to land, somewhat shaken by the experience, impressed by the strength of the water, and I sat down on the drier sands at a safe distance, convinced that I was destined to spend a few more years in the taste of life.

They are invariably noisy, those waves, dashing on and on with never-ending persistence. They have been doing this day and night, summer and winter, since time immemorial: long before humanity emerged on our planet. They don't care if people are wading or swimming near where they kiss the land, they don't care if little creatures make a living here and there in the waters. It matters nothing to them if their fury is recognized, heard, commented upon, or measured. Waves are the heart-throbs of the grand oceans all over the earth.

Some people have speculated that perhaps there is a meaning to this: "Maybe ... Sea and land meet, water and sand mingle, becoming one for a moment before resuming their individual natures. Maybe it is a song of love, of love that is older than humankind, of unity from duality. If sea and shore never met, there would be no song. Perhaps that is the mystery. We listen serenely to the song that is always different, and always the same."

An experience in mid-ocean

97% of the water on our planet is in the oceans. When we drive miles and miles across a continent, we are often impressed by the vast lands that seem endless. But the solid stretches on which we tread are small compared to the waters and ice that form much of the earth's surface. I recall voyaging from Mumbai to Genoa. The ship sailed across the Arabian sea day in and day out. I surveyed the waters on every side. It seemed like an endless stretch. One night I stood all alone on the open deck, stared at Orion and Gemini, and then at the dark sea all around. I felt as if I was all alone in a strange universe of sea and sky. The ocean was overpowering, embracing everything around."

Enormity of oceans

The Pacific Ocean alone is more expansive than all the land areas put together. This is fifteen times the land area of the United States, covering some 28% of the earth's surface. The shoreline of this grandest ocean is more than 135,000 km. It was this immense ocean with more than 4% salt content that prompted Coleridge to write the famous like: *water, water, everywhere, nor any drop to drink.* Cyclones, monsoons, typhoons, and hurricanes rage all over the oceans, and sometimes invade lands also.

If we consider the life forms that thrive therein, all our *terra firma* shrinks to insignificance. Upon, within, and deep underneath the seas, live countless creatures of all sorts of shape and size: from the tiny plankton that are tossed by the heaves of the seas and the more than twenty thousand varieties of salt-water fish to giant whales and the benthos of the nether world which are strange beings that have been living and dying for eons in the darkness of the ocean depths. Snails and sponges, starfish and seaweed, and a thousand other variety of life forms have evolved in the aquatic haven of planet earth. Indeed, our own origins have been traced to the primal soup where complex molecules, instigated by warmth and lightning and heaven knows what other stimulants, combined to form the first palpitations of terrestrial life, leading eventually to what we regard as the glorious climax of all that live: ourselves.

Having lived for so long on the solid crust, we have all but forgotten our pristine home: only the remarkable parallels in the proportions of elements in sea water and in our own bodies remind us of our aquatic origins. We sail the seas and fly across them, we surf on their tumultuous waves and we dive deep into them. Now we have become creatures that explore, explain, and also exploit our pristine home, and pollute it with lots of rubbish.

We extract elements like bromine and precious metals like zircon and gold from the sea. Years ago I visited a phosphorus mine in the Caribbean island of Curaçao and learned that the entire mountain rich in phosphorus and nitrates arose from the dried droppings of petrels which had been feeding on the fish for eons; they now serve us as fertilizers for our agriculture. To think that our lands are made to yield more crops by using the dropping of birds that had been feeding on sea-water fish millions of years ago, is as much a revelation on the interconnectedness in the world as one may receive from mystical meditation.

We get our table salt from the oceans which contain vast amounts of sodium chloride. It has been estimated that if all the salt in the oceans were spread over our lands it could form a layer 500 feet thick. When a cubic foot of sea water is evaporated we will be left with 2.2 pounds of salt. The same amount of water from Lake Michigan would yield less than a sixth of an ounce of salt, and we are not complaining. The oceans have been receiving their salts from a variety of sources: breaking up of the cooled igneous rocks of the earth's crust by weathering and erosion, the wearing down of mountains, decayed remains of millions of organisms, and also the stuff that rivers bring to it constantly.

Oceanography

Oceanographers chart the ups and downs of the ocean's rugged floor. They have mapped the oceans, and revealed many things about an aspect of the world with which most of us are unfamiliar. Someone contrasted an oceanographer with an astronomer in Freudian terms by saying that he is more interested in the bottom of the sea than in the behind of the moon. We cannot but be amazed by the richness and variety in the huge bodies of the seas. We have come to know about different kinds of coral reefs, about countless seaweeds, sponges, and about the rugged terrain and mountains in the bottom of the sea: There are ups and downs on the ocean floors, valleys and volcanoes, mountains whose tops are islands in mid-ocean. And we have added miles of cables and heaps of plastics and waste that make life difficult for many aquatic creatures Such knowledge makes us feel that, as a biological species, we are one among many, and of fairly recent vintage too. The sea creatures feed on one another, for like the law by which death is inevitable to all that is born, some creatures are destined to be fodder for others.

The Sea in poetry

     Behold the Sea,
     The opaline, the plentiful and strong,
     Yet beautiful as the rose in June,
     Fresh as the trickling rainbow in July;
     Sea full of food, and nourisher of kinds,
     Purger of earth, and medicine of men;
     Creating a sweet climate by my breath,
     Washing our harms and griefs from memory,
     And, in my mathematic ebb and flow,
     Giving a hint of that which changes not.
     (R. W. Emerson: *Sea-Shore*)
     My soul if full of longing
        For the secret of the sea,
     And the heart of the great ocean
        Sends a thrilling pulse through me.
     (Longfellow: *The Secret of the Sea*)\

What are the wild waves saying,
    Sister, the whole day long,
That ever amid our playing,
    I hear but their low, lone song?
(Joseph E. Carpenter: What are the Wild Waves Saying?

Roll on, thou deep and dark-blue Ocean, roll!
The thousand fleets sweep ove time in vain:
Man marks the earth with ruin, his control
Steps with the shore; upon the water plain
The wrecks are all thy deed, nor doth remain.
(Lord Byron: *Childe Harold*)

    A dark
Illimitable ocean without bound,
Without dimension, where length, breadth, and height
And time and place are lost.
(John Milton: *Paradise Lost*)

3
FLOATING WATER: CLOUDS

Introduction

We have all, at some time in the morning or at noon, in the afternoon or evening, or at night, looked up at the sky and watched the clouds roll by. They are shaped as heaps of cotton or as layers of white carpets, wool, as curly hair white or dark and rain-bearing. Sometimes they are moving ever so softly, peacefully floating as it were in ethereal emptiness. Sometimes they are playing hide and seek with the moon at night. Actually these sound more technical if we describe them with Latin-derived epithets. So they have been named cumulus (heap), stratus (spread out), cirrus (curl), nimbus (cloud), and so on. Those spongy masses through which our planes sometimes zoom are formed at various heights, anywhere from six thousand to twenty thousand feet above ground.

What are clouds are, and cloud-seeding

Clouds are formed when air becomes supersaturated, forcing vapor in the atmosphere to condense into visible liquid droplets. This gives rise to ice crystals in regions of the cloud where temperatures are below freezing. It is possible for liquid droplets to exist in subfreezing environments.

Clouds tell us that it is not just in the wide, wide sea, or in rivers and lakes that water abounds, but also in the earth's atmosphere. Water is present as humidity everywhere, and as clouds in the skies. The water content in the earth's atmosphere is immense, compared to what is concentrated in the clouds.

Mid-level clouds may be as high as at 20,000 feet, consisting largely of water droplets and ice crystals. High level clouds, rising still higher, are almost entirely made up of ice crystals. They are generally thin and whitish.

The stupendous amounts of condensed water don't stay there indefinitely. They precipitate as rain, replenishing pools, puddles, ponds, lakes and rivers and oceans again. We get clouds because there are extremely tiny particles in the atmosphere around which water droplets are formed. We realize that but for the billions of dust particles, smoke particles, and other tiny concentrations of matter in the atmosphere, which serve as centers around which water droplets in the clouds are formed, there would be no rain water. A perfectly clean atmosphere, with only oxygen and nitrogen gas in it would be incapable of forming clouds.

Cloud seeding

Pure water droplets can remain as liquid down to temperatures near -40°F. Clouds are agglomerations of droplets of water. When it gets to be too cold, they harbor ice particles and snow. This *supercooled liquid water* or SLW as it is called, is the key to cloud seeding. This is a technique by which one tries to divert, collapse, or disturb rain-bearing clouds in ways that would attract them to regions where there is dire need of precipitation. or make them less harmful when they deliver horrendous downpours. In the 1960s millions of dollars were spent on these programs. According to a Pentagon report, they successfully seeded clouds over South Vietnam to prolong the monsoon season so as to make it difficult for the Viet Cong to operate. There is no limit to the misuse of human knowledge.

Just as religions speak of the first rain and latter rain, cloud scientists speak of cold rain and warm rain, referring to the atmospheric conditions that provoke the rain. Since rain results from a coalescence of larger and smaller droplets, by seeding the appropriate salt crystals, one hopes to manipulate their water-delivering capacity.

Lightning and thunder

Inside cumulonimbus clouds there are both water droplets and ice crystals. Sometimes strong air currents come into play within the cloud, causing the droplets and ice crystals to collide. In the process, positive and negative charges build up within the cloud, the negative ones moving downwards while the positive ones rise to the middle and upper regions. This makes positive charges build up in the ground below. In due course the resulting electrical tensions between the positive and negative regions become so strong that a powerful electrical discharge occurs. This blazing spark in the air is what we call a lightning. Lightning flashes could come about within a cloud, between clouds, or between a cloud and the ground. In the last instance, air can get heated to anywhere from 30,000 to 50,000 degrees - which is hotter than the sun's surface: This provokes a sudden expansion of the air in the vicinity, which results in shock waves with sonorous sound. That is what we call thunder. Thus, we owe the spectacular, frightening, and sometimes devastating phenomenon of lightning and thunder to water: more exactly, to the fact that water molecules are polar.

It is estimated that some 1800 thunderstorms of varying intensities are continually occurring somewhere or another in the earth's atmosphere, and about 100 lightning discharges are flashed by Mother Nature every second of the day. The amount of electrical energy released from all this is incredibly stupendous. But we have not yet learned to harness all the electrical energy that is wasted away thus.

Because clouds can shut off light from the sun, they are sometimes used as a metaphor for obscuring, and sometimes even for gloom. Yet, one also speaks of a silver lining in a cloud, as hope even in the most depressing times, though, as Don Marquis said, "Every cloud has a silver lining, but it is sometimes difficult to get it to the mint."

Mythologies tend to view thunder and lightning as loud and awesome expressions of the wrath of

God. This reminds one of Lou Gottlieb, a folk-singer in Santa Rosa, CA who deeded his 31-acre ranch to God. When they said who would pay the estate tax, he said he would do it, for God's sake. The country reluctantly accepted. A week later, a woman in the same county hit God with a civil summons for causing a lightning bolt to strike her home. Her several hundred thousand damage suit charged God with careless and negligent operation of the universe. Her attorney said he would try to collect by attaching Gottlieb's ranch - which was a property owned by God - when and if God fails to show up in court.

Cloud forests

Figuratively speaking we talk about walking on clouds. But there are forests on which clouds walk. These are rain forests in mountainous regions which are much cooler than the low-land areas above which they rise. Here plants and trees are fed directly by the clouds, instead of by rain and rivers.

Ecologists describe them as unique eco-systems where there is an abundant vegetation such as is found nowhere else on earth. Some of these are known to have important medicinal properties. The cloud forest around Monteverde in Costa Rica is reported to be home to more than 800 species of epiphytes and 450 types of orchids. Who knows how many as yet undiscovered plants are tucked away in those regions, plants which may contain precious molecules for the cure of some of the diseases afflicting humankind! Aside from their potential value for us, cloud-forests are among the natural wonders of nature.

It was reported in the 1990s that cattle industry and coffee plantations were encroaching into cloud forests, that we intrude into rainforests and cloud forests because of our hunger for hamburger and thirst for coffee, both regular and decaffeinated. Who could have thought a hundred years ago coffee and cow-meat would be upsetting the rich abundance flora and fauna that have been thriving for ages in the warm bosom of clouds in their mountainous seclusion! Technology and human appetite know no bounds.

Clouds in Literature

The Greek playwright Aristophanes wrote a play called *The Clouds*. It is about a man called Strespiades who in his old age got into debts he could not play. He went to the philosopher Socrates to learn how to use logic for false reasoning so he could dupe the judge. It is a satire on Socrates, but has little to do with any cloud. The Bible has references to clouds.

The Sanskrit poet Kalidasa used cloud in the title of a play *Megadûtam*: *The Cloud* Messenger. In it a character is exiled from a kingdom, as punishment for neglecting his official duties. He is thus separated from his dear wife. Lonely in his exile, he sees a dark cloud at the commencement of the rainy season, and he sends a message to his beloved through this cloud. He addresses the cloud as a shelter to the distressed, and pleads with it to carry his lovelorn thoughts to the one he misses most. The idea of a lover talking to a cloud to convey his feelings is poetically charming, but the idea of sending a message via an inanimate entity up there is remarkable in its prescience: In the modern age we accomplish something similar in our telecommunication systems through microwaves that bounce off artificial satellites which are high above even where the clouds are.

Explicit in Aristophanes is the nebulous airiness of clouds. Implicit in Kalidasa is the intrinsic capacity for echo-generation in clouds.

In more modern times poets from Alexander Pope and John Milton to Shelly and Wordsworth have included clouds in their poems. Here, for example are some lines from them;

We often praise the evening clouds,
    And tints so gay and bold,
But seldom think upon our God
    Who tinges these clouds with gold.

(Walter Scott: *On the Setting Sun*)

The sun is set; and in its latest beams
You little cloud of ashen gray and gold,
Slowly upon the amber air unrolled,
The falling mantle of the Prophet seems.
(Longfellow,:*A Summer Day by the Sea*)
Were I a cloud I'd gather
    My skirts up in the air,
And fly I well know wither
    And rest I well know where.
(Robert Bridges: *Elegy: The Cliff Top*)

The clouds that gather round the setting sun
Do take a sober coloring from an eye
That hath kept watch o'er man's mortality.
(Wordsworth: *Intimations of Immortality*)

Though outwardly a gloomy shroud,
The inner half of every cloud
    Is bright and shining:

I therefore turn my clouds about
And always wear them in and out
    To show the lining.
(Ellen Thorneycoft Fowler: The *Wisdom of Folly*)

4
FLOWING WATER: RIVERS

Variety of rivers

When we see people of another ethnic background they may all appear to be the same. But when get to know them, we realize that each person is unique in many ways, and quite interesting too. The same may be said of rivers. Every river has its own personality, its own story to tell. The Ganga is different from the Godavari, the Hudson is different from the Mississippi, the Seine is different from the Rhein, the Volga from the Danube, the Nile from the Amazon: one can go on and on.

Rivers differ in depth and length, in the curvature of their course, in the ecosystems they nurture, and of course in their history. Rivers may be slow or fast, broad or narrow. They crisscross lands all over the world. Their steady flow reminds us that nothing is still and unchanging, that it is movement that keeps the world going. Recall the words of Heraclius: *panda rhei, oden menei*: Everything is in flux; nothing stands still.

Hydrological cycle

Rivers are important for a hundred reasons. Most of all, they carry fresh water that has been distilled from the salt solution of the seas through the gigantic apparatus of the earth's weather system which is fueled by solar power. Rivers are the arteries through which fresh waters flow: the water that we all need.

Rivers participate in the gigantic hydrological cycle which is one of the grand cycles in the throbs of nature's life. It is through them that the waters that rise from the oceans and precipitate back to earth return to their oceanic source. There is a beautiful prayer in the Hindu tradition which says:

>akását patitantôyam yadá gaccadi ságaram
>sarvadeva namaskárah sri kesavam
>pradigaccadi
>As the waters falling from the skies
>Return to the self-same sea
>Prostrations to god by different names
>Reach the same Divinity.

It takes a long time for nature's cycles to be completed. It has been estimated that the surface water that evaporates from the Atlantic Ocean may take ten full years before it returns to its source. In the case of the Pacific, the period is estimated to be 25 years.

In a strange sort of way, there is also a water cycle in our bodies. We drink water every day, and we also eliminate water every day. The water we drink is recycled water. When we relieve ourselves of water after retaining a portion for our internal use, that too is recycled again. In our case, the recycling is facilitated, not by the sun, but by our kidneys. Waters from all living organisms are part of this cycle. Someone once calculated that it takes some two million years for the water in one organism to go into another. I will confess I have no idea how one arrived at this figure, but it is interesting if only because what we drink today is not water that has been recently eliminated by someone else.

Direction of rivers

We have all heard of Newton's insight that it is gravity that makes the apple fall and the moon go round the earth, and the planets orbit the sun. It is the same universal gravity that causes the flow of the ribbons of water we call rivers. That's why rivers never flow upstream, even in the northern hemisphere. That's why we have in nature only water-falls, not water-rises. Of course, human ingenuity can make water rise with the help of pumps, and when you stop to think about it, this is no ordinary feat. Rivers invariably descend from higher to lower regions, often merging into large bodies of waters like lakes and oceans.

Sometimes they also connect lakes. In North America, for example, practically all the Great Lakes are interconnected by rivers. The waters of the world remind us of how connected the world is. In the story called *Paddle to the Sea* a little Indian boy who could not afford to travel all over the world, made a little canoe, inscribing on it *Paddle to the Sea*, and let it go in a lake in the mid-west. The waters take the toy boat through the various lakes and even across the ocean.

Rivers, civilizations and creatures

Rivers nourish countless life-forms. They also nurture human societies which have grown and prospered. The Greek historian Herodotus said that Egypt was a gift of the Nile. This is just one instance of the role of rivers in human civilization. The Yang-tze and the Ganga, the Amazon, the Tigris, the Euphrates, and many more have instigated great civilizations. Consider the small Connecticut river which derives its name from an Amerindian word meaning "long tidal river," because it is a long estuary. This river served as a main avenue of transportation and commerce in the early years of the colonies, and as a convenient route to the sea. With its fertile flood plains it also enabled abundant agriculture. Its water power is said to have played no small role in accelerating the Industrial Revolution, and the emergence of many towns on its banks. The steamboat was invented on this river. It was from this river that the first submarine was launched in the world.

It is not just human beings who have been taking advantage of rivers. A tremendous variety of life forms depends on rivers for their existence. To cite but one example, muskrats of the wetland ecosystem often live close to rivers. They build nice little homes for themselves, which, like the meeting place of Masons, are called lodges. The engineering skills of some of these creatures are amazing. They can swim underwater for fifteen minutes which should put any Olympic swimmer to shame.

Then there are leaping frogs, creeping snakes, slow-moving turtles and more that live in and out of rivers, like we do from our houses. Just as we sometimes get into water for a swim, these creatures step out on land for some fresh air, perhaps. Countless other animals, plants and microorganisms thrive permanently in rivers. We entice some of them with baits for our culinary satisfactions.

It is difficult to think of rivers without considering the fantastic ecosystems associated with them. Plants and trees complement rivers in serving riverine creatures in direct and indirect ways. By their shades they keep the water temperatures from getting too hot; leaves from trees fall into the river and serve as food material for aquatic life. Twigs, dead tree branches, and the roots of some trees near river banks are used by some fish as hiding places from predators. Here is a fine example of the living and the non-living combining their resources for the benefit of creatures.

Even as we do our daily chores, even as we work, play, and rest, there are processes, mighty and molecular, that keep the physical world functioning to allow for life to persist on our planet. For this we need to be grateful.

Human intrusions

In some places there are low-lying lands near the path of a river. We call them *flood-plains* because they are periodically flooded when water level rises. When the waters recede, the river leaves behind soil that is rich in nutrients for plants. For example, after the 1993 Mississippi floods receded, there was an abundant growth of plants and grass near the river banks. Floods are also helpful for the growth of fish. It has been found that the number of fish increases considerably in many rivers in years when there has been a flood.

A theologian once said facetiously that God made Man on the last day because if Man had been created on the first day, he would have started advising God as to how to make Creation even better. And God, like most seniors, doesn't like to be told what to do. Since this opportunity wasn't given to Man, one of the things we have been doing is the next best thing we could: alter every aspect of nature that we can lay our hands and minds on. So human beings have been interfering with river flows since ancient times.

As in international politics, the goal has always been self-serving, such as: avoiding flooding and ensuring a steady supply of water throughout the year. Thus, we have been straightening out the paths of meandering rivers. We have made rivers wider or deeper than how nature had fashioned them, so we can manipulate the flow to our advantage. We build levees and dikes to contain the floods. We re-channel the paths of rivers for irrigation. We build dams to prevent flooding, for storing water, for maintaining a steady stream, or for generating electricity. We don't realize that by these acts we maim and mutilate the channels that nature had carved.

Like self-interest-based foreign policy which has little consideration for its impact on other peoples, our interventions into rivers can also have disastrous effects on ecosystems that have emerged from eons of experimentation by nature. As a consequence, like refugees who are victims of wars, creatures living on shore banks are rudely displaced. By removing trees and plants so that we may build more shopping malls and condos, we facilitate soil erosion. The faster rate of flow resulting from tampering with rivers can be terribly destructive to some fish and other aquatic creatures which have been genetically or instinctively conditioned over the ages to follow certain paths. Similarly, slowing down a river is not helpful to the larvae of certain flies which are not too bright to adjust their life-cycles to changing patterns. When a river is widened and deepened, rain water tends to flush in pesticides, fertilizers, topsoil, and animal waste into it. This is like the plumbing in our toilets leading right into our living room. Not very pleasant. Needless to say, such interferences are harmful to the plants and animals which have made the rivers their homes. The topsoil we sometimes spill callously into a river tends to obscure the river. As a result, much less sunlight reaches the river-bed. This is like erecting permanent clouds over our towns. We have been doing many things to rivers that are definitely harmful to many plants, that force aquatic life to be displaced, and so on. When some townships are built, flood plains suffer severe damage whenever another flood comes into play. But thanks to a better understanding of these, these are being slowed down in some places.

Erosion

As rivers flow into the oceans, they carry with them material which are being scraped out of land. This is what land erosion is all about. This is a very complex process, depending on a variety of factors. It has been estimated that the rate at which land is disappearing due to erosion is about 2.5 inches in a thousand years. But in the very long run, the effect could be significant. Let us consider one possible long-range effect of land erosion on the United States. The average height of this land above sea-level is some 2600 feet. This means that in about 12 million years the entire United States will be eroded away! This is true of Canada too, so Americans can't just migrate to the land of their northern neighbors to save themselves before the Atlantic and the Pacific merge to become one. This is one reason why some insurance companies advise their clients that it is not financially sound to insure their homes for a period longer than four or five million years. In any case, this is not something we need to worry about as much as about the possible outcomes of the next elections.

It is an awesome thought that little by little our rivers: the very lifeblood of all civilizations, are slowly eating away the land on which we live. There is not a blessing that is without a blemish. Yet, this understanding tells us that it is silly to worry about what would happen to us when the sun extinguishes itself in another four billion years. Land erosion will not be a catastrophe for aquatic creatures, which will be the only type of creatures there may be, if at all, a billion years from now.

We also learn from rivers that the directions we take can be very significant to what happens in the long run. Recall the story of the man who said: "My grandmother started walking two miles a day when she was sixty. Now she is eighty, but we don't know where she is." In the Canadian Rockies there is a stream called Divide Creek. At a point in its course the creek divides around a large boulder. Waters which flow to the left of the boulder rush on into Kicking Horse River and finally into the Pacific Ocean. Waters which travel to the right go into the Bow River which courses into the Saskatchewan and on to Lake Winnipeg, the Nelson River, Hudson Bay, and finally reach the Atlantic Ocean. Once the waters divide on the rock, there is no turning back, and they become a continent apart.

Poems on River
    The river nobly foams and flows
    The charm of this enchanted ground,
    And all its thousand turns disclose
    Some fresher beauty varying round;
    The haughtiest breast its wish might bound
    Through life to dwell delighted here;
    Nor could on earth a spot be found
    To nature and to me so dear.
    Thus the Seer with vision clear
    Sees forms appear and disappear.
    In the perpetual round of strange,
    Mysterious change
    From birth to death, from death to birth
    From earth to heaven, from heaven to earth;
    Till glimpses more sublime
    Of things, unseen before,
    Unto his wondering eyes reveal
    The Universe, the universe as
    an immeasurable wheel
    Turning for evermore
    In the rapid and rushing river of Time.
    (Lord Byron, *The Castle crag of Drachenfels*)

Here we all work 'long the Mississippi
Here we all work while the white folk play
 Pulling' them boats from the dawn till sunset
Getting no rest till the judgement day
Don't look up and don't look down
You don't dare make the white boss frown
Bend your knees and bow your head
And pull that rope until you're dead
Let me go 'way from the Mississippi
Let me go 'way from the white man boss
Show me that stream called the River Jordan
That's the old stream that I long to cross
Old Man River, that Old Man River
He must know something, but he don't say nothing
He just keeps rolling, he keeps on rolling along
He don't plant taters, and he don't plant cotton
And them what plants 'em is soon forgotten
But Old Man River, jest keeps rolling along
You and me, we sweat and strain
Bodies all aching and wracked with pain
Tote that barge and lift that bale
You get a little drunk and you land in jail

I get weary and so sick of trying
I'm tired of living, but I'm feared of dying
And Old Man River, he just keeps rolling along
(Oscar Hammerstein, *Old Man River*)

And see the rivers how they run
Through wood and mead, in shade and sun,
Sometimes swift, sometimes slow,
Wave succeeding wave, they go
A various journey to the deep.
Like human life to endless sleep.
(John Dyer, *Grongar Hill*)

5
RAIN

Introduction

Vast amounts of water rise above the seas, only to fall back as rain. But of the all the water that descends to earth, barely ten percent falls back on land. This falling water may trickle as a light drizzle or pour down as a gentle rain, or it may come with a fury like a monsoon downpour or as a hurricane.

Sometimes when there is a downpour in the sweltering summer heat, perhaps accompanied by roaring thunder and flashing lightning, people jump with joy and run out and dance in the rain, though not with the ease and grace of a Gene Kelly. People enjoy such rain because it refreshing and cooling.

Sometimes we hear announcements of hurricane heading towards our region in the world. .We become very concerned, and many do more than their usual appeal to the Almighty to do the necessary to change the course of the erratic rain-bearing depression away from its predicted path. Often the prayers work seem to work, sometimes they simply don't, for prayers are often like playing on the slot machine: they pays back only occasionally to our satisfaction.

Value of Rain

Aside from such periodic intemperate exuberance from winds and clouds, rains are by and large beneficial to humankind. Like the rivers which by their silent course fertilize the land and serve us in many other ways, rains play an important role in feeding us all on a regular basis.

Where it rains but little, lands become parched and barren, and where it never rains, vast regions have become desert-land where it is difficult to live. It contrasts starkly from the sand on any fine beach. It is the same substance, silicon dioxide: endless strips of them adorn the fringes of oceans all over the world, vast stretches of them are laid in arid deserts, inert and desolate and subject only to the furious winds that toss them, and the glaring sun that fries them as it were. Sands remind us more than anything else that contact with water or lack of it can make all the difference in the world.

Rains have been pouring down since time immemorial. We can explain the origin and diversity of life forms in the framework of evolution, but it is difficult to understand why the forces of nature should have set up a hydrological cycle thanks to which we are able to enjoy the fruits of plants and trees, and the grains of the land.

Predictably unpredictable

Unlike sunrise and sunset, and the phases of the moon, rain is not precisely predictable. Unlike sunrise and sunset again, it is also non-uniform in where it pours. There are places in the world where there may be twenty to a hundred inches of rain in the course of a year, and in others there is hardly any rain at all. At one extreme we have Mt. Waialeale in Hawaii which averages some 450 inches a year while Iquique in Chile did not get a drop of rain for fourteen long years, and consequently no mosquitoes either. I most other places where have rains in some seasons of the year.

Rain water replenishes rivers and lakes and aquifers, besides providing enough water for plants and animals. It has been calculated that the amount of rainwater falling on the landmass of the U.S. is more than 1400 cubic miles of fresh water, and weighs over 6.6 billion tons.

Using the gift

To what extent one utilizes a gift depends on the circumstances under which the recipient functions. Children in a family inheriting equal shares of patrimony utilize, spend, or dissipate what they get in different ways. So it is with what happens with rainwater too. Depending on the rate of rainfall, on the terrain, on soil conditions, on the density of vegetation, on urban conditions and other such factors, the water that falls on the ground as rain may flow into a river or a lake or seep underground. When rain falls in a hilly region it streams off downhill, and finds a place in a lake or a pond. If there is much clay in the soil it is more difficult for the water to soak into. With more sandy soil the situation is different. Lots of vegetation slows down the flow and thereby land erosion is also much less. Roads and pavements in cities are not helpful for natural runoff of rain water.

The variety of ways in which rainwater serves us is endless. Irrigation is of course rain water's first use. But we also need it for cleaning fruits and vegetables, and for washing everything from clothes and cars to cups and dishes. We rarely think of it in the context of steel making or for cooling equipment in power plants: whether thermal or nuclear, but here too we utilize enormous amounts of water. Heavy rains support our rain-forests which cover only 7% of the earth's surface, yet nearly one half of the world's species of plants and animals are found there.

The raindrops which keep falling on our heads are usually spherical in shape and about a millimeter in radius. Sometimes they may become almost five times as large, but then, because of air pressure and surface tension they get distorted and break up.

Odyssey of Water Molecules

If we wish to get an idea of how stupendously complex the physical world is, consider the possibilities that are open to a bunch of water molecules in the ocean. As a result of forces beyond their control they may come to the surface of the sea. There, as a result of exposure to solar radiation they absorb enough energy to break away from the bonds of fellow molecules, and in their freedom from bonds they rise up as water vapor. The winds thrust them into the upper layers of the atmosphere where they cool down and become water again, or if they move still higher, they could become ice crystals, often around tiny particles of dust or smoke. These droplets wander in clusters (clouds) in the air and eventually fall down due to gravity.

Now they might flow back on another section of the ocean waters, for they could have been transported by the winds miles away from where they ascended. Or gain, they might tumble down to land, and now they have a wide range of possibilities. They could land on a leaf or become dew. They could fall into the bosom of a lake or river. They could seep into the ground. The rainwater that trickles underground recharges the aquifers which are depleted through wells and natural fountains. Aquifers are water-filled regions between the rocks underground. It is common knowledge that we get oil by digging. But we also get water by digging. In a good many places in the world, people rely on well water for their subsistence, as do the roots of trees and many plants everywhere.

Or they could run into a sewer system, or be gulped by an avid rain-bather who stuck out her tongue in the downpour. The possibilities are endless and unimaginably vast.

If such are the unpredictable courses for a motley group of water molecules, what can we say about what might happen to each of us in a lifetime and beyond! Like the molecules which are blissfully unaware of what nature has in store for them, we too may be just drifting at a level of consciousness that seems to be the only reality we know.

Rainwater Pollution

In our own times, the pristine purity of distilled rain water has become a fantasy: a thing of a bygone age of industrial innocence. How ironic that after science and technology channeled clean and purified water into our homes through plumbing and faucets, waters the clouds are sometimes acidic solutions For all the oxides of sulfur and nitrogen that have become part of civilization's exhalation have transformed summer showers into acid rain. They lower the pH levels in our lakes, making life difficult for fish. Thanks to our factories and industries, paper mils and furnaces, we have been slowly producing the poisons in our rains.

It says in the Bhagavad Gita: "From grains arise all creatures, And grains emerge from rains." From the simplest to the most complex human activity, from the meanest to the noblest, from the basic biological to the most sublime: intellectual, moral, spiritual or whatever, consider all that go to make society and civilization rich and worthy: the arts, music, literature, science, sports and everything there is. Ultimately what sustains all this is the food we eat. Deprive any community of food for a week, and it will degenerate into savagery and perish. As Pliny the Younger stated, "the body must be repaired and supported, if we would preserve the mind in all its vigor." Yet, when we talk of human activity, we rarely refer to food as the most fundamental of everything, and seldom mention water.

Rainbow

The magnificent multicolored tenuous rainbow that we sometimes see in the sky after it has rained is essentially due to a bunch of water drops in the atmosphere which refract sunlight of various wavelengths internally, and splash them out as the semi-circular arc spanning the heavens. No paint or pigment, pen or pencil, can draw a picture stretching from ground to ground, curving all the way to the distant sky, as water droplets hanging in the air do. Incredibly they only use the natural light from the sun.

Poems on Rain

      The thirsty earth soaks up the rain,
      Ans drinks and gapes up for drink again;
      The plants suck up in the earth, and are
      With constant drinking fresh and fair.
      (Anacreon, *Odes*.)

How it pours, pours, pours,
    In a never-ending sheet!
How it drives beneath the floors
    How it soaks the passer's feet!

How it rattle on the shutter!
    How it rumples in the lawn!
How 'will sigh, and moan, and mutter,
    From darkness until dawn.
(Rossiter Johnson, *Rhyme of the Rain*.)

The day is cold an dark and dreary;
It rains, and the wind is never weary;
The vine still clings to the moldering wall,
But at every gust, the dead leaves fall,
And the day is dark and dreary.
(Longfellow, *Travels by the fireside*.)

The rainbow never tells me
That gust and storm are by;
Yet she is more convincing
Than philosophy.
(Emily Dickinson, *Further Poems*, 48)

6
FROZEN WATER: ICE

WATER AS ICE: FROZEN WATER

Introduction
When we say H-2-O we immediately think of water. But we shouldn't forget that ice is also H-2-O. When we say ice, we think of it as little cubes, sometimes with a hole in the middle: blobs in a glass to cool whiskey or soda; or as blocks wrapped in a towel to soothe the pain from a swelling.

Snowflakes
There are also tiny bits of ice, soft and easily molten by the mere touch of a hand. We have all watched them with rapture as they fall ever so prettily on a wintry morn, landing sometimes on the branches of trees. Yes, these are the wondrous snowflakes which too are made up of ice: i.e. out of very cold water. Who but Nature can chisel such wondrous crystals from tiny bits of water! So soft and delicate, and beautiful too, though fleeting in

their existence, as they gently come down from up above. As Mary Mapes Dodge put it poetically:

>Whenever a snowflake leaves the sky,
>It turns and turns to say "Good-bye!"
>"Good-bye, dear clouds, so cool and gray!"
>Then lightly travels on its way.
>But when a snow-flake brave and meek,
>Lights on a rosy maiden's cheek,
>It says, "How warm and soft the day!"
>"'Tis summer!" and it melts away.

Snow

Unlike the little droplets of water that make up the rain, snowflakes are made up of tiny crystals. A crystal is a miniature sculpture of atoms and molecules constructed by nature, for in a crystal the molecules are arranged in patterned lattices in geometric order. Solid state physics has studied them extensively, and classified crystals in terms of their geometry. In snowflakes the molecules form hexagonal lattices of oxygen and hydrogen atoms. The aesthetic dimension of natural order is often as dramatic and impressive as the mathematical precision implicit in it. The crystalline form gives ice its magic.

Snow crystals are different from raindrops that freeze in extreme cold, creating sleet which is just a slippery stretch that is good for skating and bad for walking or driving. This reminds me of the man whose car wouldn't move because the driveway was covered with snow and sleet. He called the garage and they brought him snow tires. When the mechanic came to replace the rear wheels, the man said, "No, no, we need to replace the front tires. The rear wheels are turning perfectly. The front ones are the ones that are not moving."

One may wonder why, whereas water is colorless, snow is white? Actually, the individual crystals are transparent. It is the play of sunlight as it is reflected internally and escapes out with all its component colors that makes the snow appear glaringly white. The wonders that light and matter play together are often the source of immense beauty. Similarly light-scattering effects are involved in the sea where colorless water turns beautifully blue.

Some have seen a symbolism in snow: When we are snow-bound we feel helpless, but in due course the snow melts away. So it is with the problems we face in life. However horrendous they seem at the moment they too will melt away. The French poet François Villon exclaimed, *Où sont les nièges d'antan?*: Where are the snows of yesteryear? They are nowhere to be found in the land around us when we are in the warmer seasons spring and summer. From every aspect of Nature, from every transformation that is ceaselessly occurring around us, the reflecting human mind can fish out a lesson or two of significance.

Glaciers

There are also vast sheets of ice on the surface of our planet which have lasted a million years. On the same earth where there are stupendous stretches of hot sands and dry dunes, where oceans rage with huge waves and rivers meander smoothly, there are also enormous glaciers, often in serene stillness. These are accumulations of ice and water, air and rock debris. They flow ever so slowly. Such immense ice may be vast as a continent or fill a valley between two large mountains. Expansive spans of ice have spread and shrunk over the ages, at least twenty different times, geophysicists say. We have come to know much about the glacial components of our environment, about ice-ages that have come and gone, and which may come again in ages yet unborn, but our current concern is with global warming.

Two processes are continually under way in glaciers: *accumulation* by which matter is added to the glacier, usually through precipitation of snow and rain; and *ablation*, by which matter is lost through melting and evaporation. Not unlike human bodies which take in and give out, glaciers survive and grow when accumulation exceeds ablation. They diminish and disappear when the contrary is the case. What an extraordinary balancing act nature is performing, where sometimes one side wins, sometimes the other!

One may be inclined to think of glaciers as existing only in the temperate zones of the planet, and confined to the polar zones. Not really. They are not only in Greenland and Antarctica, but also in Chile and New Zealand, some in the mountainous regions in the Himalayas in India and some near equatorial Africa. Parts of them melt away in summer, but they regain their lost substance when winter sets in. The largest volume of ice accumulation and ablation occurs in temperate glacial systems because of very high precipitation. There is no limit to the wonders on our planet, to the extremes of temperatures it supports, and to the organisms that thrive under extreme conditions.

Glaciers are in dynamic equilibrium with their environment. This means two things: One the one hand, when there are minor variations in the physical conditions, they tend to shift one way or another and get back to their equilibrium state. However, if the perturbations are significant, the impact on the glacier could be great. A small valley glacier may respond to adverse changes in a hundred to a thousand years, whereas the response time for an ice sheet may be anywhere from ten thousand to a hundred thousand years. Thus, if we are to list the many natural catastrophes that could spell disaster to human life, we must include glaciers: Should all glaciers melt, much of the lands on which we live and work would be submerged in an irretrievable an aquatic realm.

Glaciers melt under a variety of environmental changes: from atmospheric changes to climatic warming. Sometimes they are affected by changes in sea levels. Depending on how much water melts at the base, the flow rate will be affected. Ultimately it is the relative amounts of accumulation and ablation that determine a glacier's destiny.

The wonder of Ice

Frozen water is abundant on earth, as perhaps in other planets too. Looking at a transparent melting ice cube one can hardly suspect what its molecular structure is. Yes, it too is made up of two atoms of hydrogen and one of oxygen – the universally known H_2O. At the deepest level ice is a crystal: geometrical (hexagonally structured) groups of molecules. This type of ice is known as Ib.

Other types of ice, bearing such names as Ice II, Ice III, and Ice IV have trigonal, tetragonal, and monoclinic crystalline structures.

The nature of the bonding between its constituent bonds endows it with properties that are unique to ice. It is now believed that there can be at least seventeen different kinds of ice-crystals, though we find only one of them here on earth. Others could be elsewhere in the vast wide world.

One of the most interesting properties of ice is that it is less dense than liquid water: this is what makes ice float on water, rather than sink into it. This is a property of the utmost importance for aquatic life. For, were it not so, come freezing winter, ice sheets on lakes and ponds would move en masse to the bottom, crushing the creatures in the body of water.

When one thinks of basic research in science often high energy physics, unified field theories, and large colliders come to mind. But considerable basic research is also devoted to the study of the myriad mysteries of ice. Understanding of these properties is very relevant and essential in such fields as glaciology, dating of ice cores in Antarctica, thunderstorm electricity and comets, to name a few.

Shifting nature of planetary concerns

A few scientists were worried about the onset of an ice age. Now we are terrified about global warning. Unfortunately quite a few such predictions - whether of planetary warming or cooling, of the greenhouse effect or the all-is-fine conviction - are influenced to an intractable degree by political factors, economic motives, difficult-to-interpret data, complex possibilities, let alone gross ignorance.

In 2000 there was a front page article in the New York Times to the effect that a hole one mile in diameter had been detected in the arctic regions such as had never occurred for at least six million years. This led to speculations that we are on the road to drowning all together in the not very distant future, because the polar ice caps had begun to melt. A few months later, the Times retracted the story.

Some years ago there was a scientific paper in the prestigious journal *Nature* which suggested that global warming can sometimes lead to even a worldwide freeze. A million years ago the North American continent was covered by an ice sheet two miles thick, we are told. What a chilling thought that is! And when the earth began to warm up some 10,000 years ago, that sheet retracted to the polar regions. The ice sheets left in its wake at least two lakes containing more water than the Great Lakes combined. To quote from them, "In the Hudson Bay, ice held the water in place like a plug in a bathtub. When the plug finally melted, trillions of gallons gushed into the Labrador Sea, flowing out at 100 times the rate water leaves the Mississippi." They went on to say that if a modern glacier such as the Greenland Ice Sheet melts as a result of rising temperatures in the next century, it could trigger a similar flood and climate fluctuation.

Records show tha from 1860 to 1940, the average temperature of the globe increased by about 0.4 degrees C. During the next forty years it cooled down by 0.1 degree. Then it has warmed again by 0.3 degree during the past 30 years or so.

Poems on Ice
>   Three children sliding on the ice,
>   Upon a summer day,
>   As it fell out, they all fell in,
>   The rest they ran away.
>   (*The Lamentation of a Bad Market.*)

>   Some say the world will end in fire,
>   Some say in ice.
>   From what I've tasted of desire

I hold with those who favor fire.
But if it had to perish twice,
I think I know enough of hate
To say that for destruction ice
Is also great
And would suffice.
(*Robert Frost, Fire and Ice.*)

Be thou as chaste as chaste as ice,
as pure as snow,
hou shalt not escape calumny.
    (Wm Shakespeare, *Hamlet*.)

7
HINDUISM

Water and the Hindu tradition

The Himalayas are mountains are the tallest in the world. They have, since remote ages, evoked awe and reverence in the hearts of the Indian people. Its perennially snow-capped peaks have been described as penetrating into the very mysteries of heaven. Thence originate the rivers that water the plains and make them agriculturally abundant. Secluded spots in the Himalayan region, near a pond or a lake, a stream or a river, have always been deemed appropriate for meditation on the divine.

In the Hindu world, rivers are regarded as sacred, and the Ganges (Ganga) the most sacred of them all. It has its origins in the Himalayan peaks, and as per Hindu mythopoesy, it trickles through the matted hair of Lord Shiva who is seated up there with his consort Párvati in serene meditation. The Ganga is seen as a descent form the galactic strip that spans the heavens, and has come here below to sanctify the earth. When one considers what it does to the Indo-Gangetic plain, transforming the vast land into fertile soil, prompting cities and shrines along its path. If that is sacred which evokes in us a feeling of reverence by virtue of its power and influence on the human condition, then every river is sacred, and certainly the Ganga.

Rishikesh

Among the many sanctified spots on its long course is the little town of Rishikesh which sits quietly on the banks of the sacred Ganga not far from its origins before the river sanctifies other centers in the Hindu world, peacefully flowing at one place and furiously gushing at another. Here you find short walkways flanked by multicolored icons of personages and episodes from Hindu mythology and tradition.

As generations come and go, grandparents bidding farewell to grandchildren, it is not simply the genes that are carried through, but the intangible sounds and gestures of rituals too, as also concepts and belief-systems that were forged millennia ago, for that is how culture is transmitted from generation to generation.

I recall that early one morning before dawn I walked to the river bank to sit there in silent meditation, only to discover that at least a score of people had come there just for that purpose before larger crowds would assemble there. Once I squatted and closed my eyes and receded to the inner world in a meditative mood, everything around vanished from my conscious grasp, and I could barely feel the presence of anything or anybody. I felt the tranquility that is promised to all who sit on the banks of the Ganga whose waters seem crystal clear, so unlike the Ganges I had known in the lower regions near Varanasi and Kolkata, for by then he river has accumulated many leavings from the teeming millions.

Water as Divinity

In the Hindu tradition, water is compared to Divinity. For the Divine, like water, is pictured as pure in essence, imperceptible by its transparency, yet pervading everywhere. As waters flow from high to low, and serves many human needs, the Divine descends from high heavens to the earth below to help us in our predicaments. Water cleanses that which is unclean and transforms it into a state of purity. So too those who incorporate the thought of the Divine are cleansed of worldly sins and purified at heart.

Pilgrimage

There is a body of water near every Hindu place of worship. It may be a river, a stream or just a little pond. Every place of pilgrimage is regarded as springing from a sacred water. Pilgrims carry a bottle of water from the sacred rivers, and bring it home for its sanctity.

There is a spot in the plains where three of the sacred rivers are said to converge: the Ganga, the Yamuna, and the submerged river Saraswati. Known as *Triveni* this is where millions gather every three years during the winter solstice, and thousands take collective dips into the sacred waters for spiritual purification.

Bodies of water, when connected, eventually come to the same level. Water is an equalizer, as it were. This is a reason why when pilgrims dunk their heads in the waters of sacred spots, all distinctions of rank and class and caste are said to disappear.

Yamuna

Yamuna is another sacred river which has its origins in Himalayan peaks. Originating from the Champasar Glacier at an altitude of almost 4.5 km, it is said that the source of the river is the glacial lake of Saptarishi Kund. There is a sacred shrine of Yamunotri near this source at an altitude of 3.2 km.

In the Hindu tradition, as in ancient Greece at one time, rivers are personified, and beautiful stories are woven around them. Yamuna is associated with the name of Krishna, the fount of wisdom who, in Hindu sacred history was the source of the Bhagavad Gita: The Bible of the Hindu world. The personified Yamuna regarded as the consort of the divine Krishna. The river is said to have gone around Krishna before coming down to earth. She descended down the hills and reached the plains at Khandav Vana which, it is said, has been developed as the city of Delhi now.

Yamuna meanders after going through Mathura, a place where Krishna once to have sported. From here it goes to Agra, which is famous for the Taj Mahal, then proceeds to Gaya, another sacred city, and finally joins the sacred Ganga. It is said that Krishna played on the banks of river Yamuna during his childhood days. Yamuna courses through Vrindavan and Mathura, cities which are also associated with Krishna. Then she flows towards south and southeastern parts, nourishing the places on her way.

Sarasvati

There is a festival in which people of the Hindu world pay homage to Sarasvati, the divine representation of learning and knowledge. The name *Sarasvati* literally means *She of the lake*. In Vedic times this was the name of a river. That rivers are the life-blood of civilizations was recognized by the people of India since ancient times. They realized that water can cleanse, so they took it as a symbol of purity. They saw the clarity of water, so they took it as a symbol of the awakened mind. They were aware of the value of fluids for the proper functioning of the body, so regarded it as medicinal. They understood that we need to cross the river to reach the opposite shore. So they took it as a symbol for what needs to be crossed to go from our present state (of ignorance) to the other state (of enlightenment). They observed the river flowing from the distant snow-capped peaks of the Himalayas. So they regarded Sarasvati as heavenly. The sage-poets who composed the Vedas invoked her for blessings. In the Rig Veda one reads:

ambitame naditame devitame sarasvati.

The best of all mothers, Most supreme of all rivers,

Most divine of all Goddesses, O Sarasvati!

Sarasvati is the fount of wisdom, the supreme symbol of all that the intellect can accomplish. for Sarasvati is the all-embracing principle implicit in every book and notebook, in every paper and blackboard. Sarasvati is present on every bookshelf, and in every library in the world. She is there in every mind that thinks, in every eloquent tongue that speaks, and in the creative genius of sublime poets from whose hearts and minds words flow like pristine torrents. And she is a river who is remembered to this day as the founder of Indic civilization, and personified as a woman of grace who holds a musical instrument.

Bless me with knowledge and with memory, Oh
Source of all Knowledge!
Bless me with perseverance and poetry,
And the ability to teach students.

That is why she is revered in places of learning, by children in schools and by professors at universities alike.

Ashes of the departed and sacred waters

According to a belief in the Hindu world if the ashes of the departed are strewn in the gushing Ganga, the departed spirit will find post-mortem peace. I recall how, many decades ago my father took the family to Varanasi: the most sacred city in the Hindu world, which stands on the Ganga. He carried with him an urn containing the cremated remains of my grandfather. It was a memorable experience: Standing on the steps of one of the many ghats - as the pilgrimage stations on the river are called - I heard the meticulous chants of the priest, repeated by my father, after which the ashes were dispersed in the flowing river. My father had discharged an important filial duty.

When we talk of meaning, we need to realize that beyond individual appraisals of what life and existence are all about, the cultures of humankind have, over the ages, spelled out meaningful actions. What matters is not whether they carry the support of science and logic, but whether they bring solace and satisfaction to the practitioners.

The waters of India, whether flowing as surging rivers or standing still in a pond at the entry of a temple, have served the human heart in magical and meaningful ways.

8
BUDDHISM AND WATER

As in other traditions water plays an important role in the Buddhist world also. There is a legend in Buddhist lore to the effect that at one time Ananda, a disciple of Buddha, put an end to a series of natural disasters like famine and pestilence by simply uttering a sacred mantra and flinging some water skyward. This is just one instance of the belief that when water is used in conjunction with sacred chanting, it can be very powerful. This idea is implicit in many religious practices.

Buddhist monks sanctify water by reciting from *paritta* - Buddhist hymns. Such water is said to have spiritual value, and also curative effects. It is believed in the tradition that water that has been sanctified by uttering the appropriate mantras acquires a spiritual potency which can infuse the person who imbibes it with wisdom. Just as the chemical molecules of water serve the physical basis of our existence, the sacred syllables utters into the water infuses it with this special property which has an effect only on the faithful. Such water can also ensure happiness: This is the reason why water is ritualistically poured into the cupped hands of bride and groom during Buddhist wedding ceremonies. The wedding guests bless the newly married couple in this way. Buddhists also use water to clean, both physically and spiritually, the body of a person who has died.

When gifts are offered, there is the giving of holy water also. This is to remind us of an event in Buddhist history in which the king Bimbisara granted a special park for a Buddhist Order with a ritual offering of water.

Water has been mentioned in Buddhist lore as a medium for testing one's faith and determination. According to a well-known story, the great Buddha was once seated in the outskirts of a village on the banks of a river. Now it happened that there was a man named Sariputta on the other side of the river who felt a deep need to go to the Buddha for wisdom. But the waters of the river were turbulent, and there was no bridge to cross. Sariputta said to himself: "This will not prevent me from going to the Blessed One." And he stepped over the running water which thereupon became firm as a granitite slab for him to walk over. Somewhere halfway, Sariputta's mind wavered, and the slabs began to melt. But when he refocused his mind and regained his faith, the water became walkable again, and he reached the other bank.

This story is told to emphasize the power of faith in achieving the impossible.

In some Buddhist countries, like Thailand, they have a special festival for the Goddess of Water. Water also plays an important role on New Year's Day. It is said that on this day water acquires special powers, and rids one of every trace of bad-luck. One prepares water with scents and offers it to cleanse.

9
CHRISTIANITY AND WATER

The sanctity of water occurs in at least three important contexts in the Christian world, each in its own way referring to a core belief of the tradition.

First there is baptism which is a sacrament of the religion: a formal procedure which has deep spiritual significance. In the spiritual framework, human beings have two births: First, there is the physical birth in which we come out of the mother's womb in whose waters we had been protected till then. This is followed by a second birth when we wake up to our spiritual dimension of life. One is said to be *reborn to Christ* when one is baptized. In this sense, one who is baptized is born again.

But there was more to baptism than the second birth. Through that sacrament one promises to be faithful to

Christ and to the Church. When a child is baptized its parents dedicate its entire life to Christ. That is why baptism is identified with Christening which implies more than giving a Christian name. When a child is christened, it is being dedicated to the service of Christ, abbreviated as IHS: In His Service. That is why when ships are christened in Great Britain they are dedicated to HMS: to His or Her Majesty's Service. The practice of christening ships arose from the fact that in earlier times, Britain, like many other nations, was under Christian laws, as many Islamic countries are governed in our own times by Islamic laws. Therefore, something as important as a ship had also to be dedicated formally to the service of Christ. In the ship ritual, however, numerous champagne bottles are broken with little stimulation to eager imbibers of the beverage.

Baptism may be compared to the formality by which one becomes citizen of a new country, for here too there is allegiance to a flag, and to the acceptance of the laws of the land. Here too there are formalities to be followed by the new citizen. There is more than an analogy here. In former times, becoming a citizen involved a promise of loyalty to the new country, dedication to its well being, and willingness to serve it in times of need. Very much like that, just as for many in our times, taking up citizenship has become more a matter of convenience of travel and job offers than any feeling of commitment to the country, in the Christian world also baptism is often reduced to a routine ritual rather than something of profound spiritual value.

Different interpretations and explanations have been given for baptism and its significance. In Mark we read that baptism is for the forgiveness of sin. In John it is for repentance. But then, one may wonder why Christ, who was himself without sin, was baptized. The answer one gives for this is that the goal in this case was to "fulfill all righteousness." That is to say, it was for a complete transformation of all of humanity to the path of righteousness.

In any event, in later times all these were combined: Baptism became a ritual for conversion, as a declaration of life-directing faith in Jesus as the Christ, and also as an expression of repentance, and the washing away of all sins. Baptism is meant to save a person's soul by placing a person in the body of Christ.

There is another symbolism for baptism which is given in Roman (6:3). Here we are told that the immersion in water and the coming out of it is symbolic of the death, burial, and resurrection of Jesus Christ. Paul says: "Know ye not, that so many of us as were baptized into Jesus Christ were baptized unto his death? Therefore we are buried with him by baptism into death: that like as Christ was raised up from the dead by the glory of the Father, even so we also should walk in newness of life. For if we have been planted together in the likeness of his death, we shall be also in the likeness of his resurrection."

The Jordan River

The Jordan River is a river whose sanctity is associated with a historical event. Here is where Christ was baptized by John. It was already a sacred river which the nation of Israel had crossed, guided by Joshua, to reach the promised land only to find more than thirty enemies, Jesus crossed over the chasm between Heaven and earth to save one and all from enemies. The imagery of crossing a river to go from one experience of life to another is not unlike the Hindu vision where the waters symbolize something that needs to be crossed in order to get a glimpse of spiritual reality. It has been said that when the priests of Israel stepped into the swift river, they found dry ground for them to stand. Likewise, when Christ waded into the waters of Jordan he established a firm foundation for one and all. This river is mentioned in many contexts in the Bible.

The importance of the River Jordan is brought out also in an incident in II Kings (5): Na'aman, the leper, was asked by Elisha, the man of God, to obey God's commandment and take a dip in the River Jordan; the Syrian was angry, and he asked why the rivers in Damascus wouldn't be just as good. His servants told him that if God had commanded him to go to the River Jordan that is where he should go. He did exactly that, and was cured. It is often said that the story is meant to stress the importance of obeying God's command. Perhaps it is also to remind us that there is a difference between water and water. In the religious/spiritual context, when talks about the sacredness of water, it is not just to the chemical compound that one refers. Rather water becomes sacred when it is associated with the divine in one form or another: It may flow from the matted hair of Lord Shiva, it may be the water that was made to part by Moses, or it may be that of the river in which Christ was baptized, or it may the water which is in a well dug by Gabriel's heels in

Makkah,. From religious perspectives - contrary to that of religious naturalists - there is no sacredness without association with something spiritual, supernatural, or mythological.

Historical note on baptism

From a historical perspective, the idea behind baptism predates Christianity. It is known that the worshippers of Isis in ancient Greece were initiated per *lavacrum*: through a font. Likewise in the mysteries of Apollo and Eleusis men were baptized "unto regeneration and exemption from the guilt of their perjuries." According to Tertullian, "Among the ancients, anyone who had stained himself with homicide went in search of waters that could purge him of his guilt."

Scholars tell us that the earliest reference to baptism as a Christian practice is found only in Justin's *Teaching of the Apostles* which is dated as between 90 and 120 of the Common Era. Here it is said that if there is no living water, one could baptize into other water. If one cannot baptize in cold water, one may do so in warm water. It is not clear what is meant by the term here. Perhaps the reference is to the River Jordan.

Living water

There is an incident in the fourth chapter of John where the idea is mentioned. Jesus is at a well, tired after coming to a town in Samaria. A woman comes there to draw some water, and Jesus asks her for a drink. She is surprised that he of the Jewish land asks her, a Samaritan for water. It would seem that already in those days neighbors were not exactly friendly: One reason why Jesus preached, "Love thy neighbor." The woman began to argue with Jesus, and Jesus says that if she knew who she was talking to, she wouldn't be talking to him like that. And here he utters the famous statement, referring to the well water: "Whosoever drinks from this water shall thirst again: But whoever drinks the water that I shall give shall never thirst again; but the water that I shall give him shall be in him a well of water springing up into everlasting life." The woman said to him, "Sir, give me this water, so that I will thirst not, or come here to draw."

In this exchange we see the dual significance of water: Water as a substance that quenches the thirst of our physical body, and water as that which quenches our spiritual needs. We are told that this other water which Jesus offers is of eternal significance. The idea is that the restlessness and longing, the deep dissatisfaction and emptiness that one feels without a spiritual framework will vanish when one accepts the gift that is brought to us from the transcendental source. This is the recurring theme in all religions.

Christ's return as the Latter Rain

Some groups in the Christian tradition interpret the word rain in the Old Testament as a metaphor for Christ. There are passages in the Bible which they take as the promise of the coming of Christ. Thus, for example, in Deuteronomy (11:13-17) we read: "And it shall come to pass ... that I will give you the rain of your land in his due season, the first rain and the latter rain, that you may gather in thy corn, and thy wine, and thy oil. And I will send grass in thy fields for thy cattle, that you may eat and be full. Take heed to yourselves, that your heart be not deceived, and you turn aside, and serve other gods, and worship them; And then the Lord's wrath be kindled against you, and he shut up the heaven, and there be no rain." In the view of some Christians, the first rain refers to the Christ who was born in Bethlehem. They give as further proof the lines from Joel (2:23) where it says: " **God...he hath given you the former rain moderately, and he will cause to come down for you the rain, the former rain, and the latter rain in the first month.**"

In this framework, it is believed that the major problems we are facing are expressions of God's wrath which is now turned against us because humanity has abandoned its spiritual quest and has begun to worship other gods such as materialism, greed, and economic prosperity. Furthermore, in their view, after a worldwide punishment for all these, Christ will come back to earth, and that is the latter rain referred to in the passage above.

10
ISLAM AND WATER

The reader can get all the relevant information on this from:
https://archive.islamonline.net/?p=17504

11
CONCLUDING THOUGHTS

Water is at the very foundation of our biological being: every living cell contains water. Even tons of coal and millions of barrels of oil would be utterly useless as sources of energy on a planet where there is no oxygen. Likewise, all the calcium and sodium, phosphorus and other minerals our cells need would be utterly unusable without water in which their ions can migrate.

A religious traditions regard water as sacred. In some traditions, water itself is worshiped, in others it is used in worship. Water is invariably regarded, not only as the cleanser of the body, but also as the purifier of the spirit. It is used as a medium in rites, and as a symbol in doctrines. It has been personified in mythologies, and deified in some religions.

Poets have sung about water as such, or about its countless presence in different forms: as lakes, rivers, and brooks; as falls, oceans, and clouds; as steam, snow, and solid ice. Water is present in all these common forms, but it is also invisible as vapor and humidity.

Water is part of our physical being: It is there in our sweat and saliva, in our blood and urine too. In moments of joy and sorrow we shed tears which are water. Even our language has been touched by water. In extreme elation we say we are walking in the clouds. In referring to the irrevocable past we speak of water under the bridge. A weak argument holds no water, and when something loses intensity we say it is watered down. When we blurt out in anger, we are said to be letting off steam. When things are unassailably secure, we call them water tight. We compare a predicament to fish out of water. We talk of spending money like water. To say no matter what, we say come hell or high water, and we speak of keeping head above water.

In very ancient times, before humankind invented mirror, it was water that revealed to people how they looked. For how else can one see oneself in a world without mirrors, if not by seeing one's reflection in a body of water? By reflecting it made us reflect. Again, it was water which made us aware of the laws of refraction of light, by observing the apparent bending of a stick half immerse in water.

Blessed as our planet is in harboring so much water, it is not unique in this regard. Theoretical astronomy had suspected, and observational astronomy has detected, water in other nooks in the universe. Spacecraft orbiting the moon, by using radar, have detected ice in certain craters near the poles. There is reason to believe that there may be large amounts of water underneath the lunar surface.

Comets are known to be chunks of dust and frozen gases and water too. Indeed, an un-poetic description of comets is that they are dirty snowballs. Some have even suggested that comets may have been the sources of the water on earth, through millions of collisions over billions of years.

The *Pathfinder Lander* sent us photographs which show stacked boulders that could have been left their by powerful floods. Since Mars has a much, much thinner atmosphere than ours -about a hundredth in density - liquid water must have evaporated from its surface in the extreme cold. There are indications of huge polar ice caps of carbon dioxide on Mars which change in size with Marian seasons. The spacecraft Galileo sent pictures of fissures on Europa, one of Jupiter's satellites, which look suspiciously as if they were formed by frozen ice.

There are what are called *Water Masers* (Microwave Amplification by the Stimulated Emission of Radiation) in interstellar clouds. The term refers to water molecules which seem to be stimulated by energies in nearby stars. Such powerful masers have also been detected near the centers of other galaxies. Planet earth may not be alone in the universe to have inherited water, and thence to have given rise to the phenomenon we call life.

Since the most ancient times men and women have lived their lives from the waters of streams and rivers, ponds and wells. Sometimes there have been long periods without rain, causing draught and famine, starvation and death. But by and large, the world never confronted the kind of water crisis that humanity faces today.

Gone are the glory days when water was within reach of everyone, when water evoked only images of the irrigation of fields, of the cleaning of body, the quenching of thirst, and sprinkling to sanctify. Gone are the days when people sang songs as they sailed on the canals of Venice, when people experienced spiritual elation when they stood in the waters of the Ganga and prayed, when the waters of the Seine gracefully moved through Paris inspiring poets and lovers, when the Mississippi was just another mighty river bringing abundance to the lands on its course, when the lakes were for carefree swimming and lazy fishing, when rains were acid-free and oceans were filth-free.

There was a time when people drank freely from a river or a well. Today millions walk around with plastic bottles containing allegedly spring water, often with a French brand name, and paying for it too. Like bringing up children and caring for grandparents, water is no longer a source of joy but a problem to be faced, a crisis to be dealt with, a threat for the future.

It is true that in a few countries people have more than one bathroom per household, several water faucets, and Jacuzzi to boot. But these are rare pleasure-gardens compared to what is available for the vast majority of humankind. Many cities all across the world have periodic water-shortage. According to a United Nations report, more than half of Europe's cities are exploiting groundwater at unsustainable rates. Chronic water shortages are affecting four and a half million people in Catalonia, where authorities are pressing for the construction of a pipeline to divert water from the Rhone in France to Barcelona. For more than a billion people, there is no access to safe drinking water. For them, a bucketful of clean water a day would be a wonderful blessing. That is why states within a country sometimes fight about sharing water from the same river, and nations guard jealously their waters when a river spills beyond their borders.

The sacred rivers of India are in crisis. The Nile and other rivers have been dammed. The wetlands of Sundarban and the forests in Bangladesh are in serious trouble. As we do more logging and build more structures, we cause the glaciers feeding the rivers to melt more rapidly.

Every informed person and group assures that water shortage will be one of the two most serious problems in the next few decades, global warming being the other.

One can go on and on, elaborating on the sad story of how the rivers that nurtured civilizations are now threatened by the very civilizations they nurtured. We can only hope that politicians and economists, scientists and engineers, and every responsible citizen of the planet will do whatever one can to alleviate the situation.

It is important that in the face of all these horrendous problems we don't succumb to despair. Rather, we should make modest efforts towards their solution in our different ways. The one lesson we learn from humanity's history is that our species has an extraordinary will and capacity to survive. True, when we read in the papers and hear in the news about those who fight and grab, about people who try to resolve problems through hate and war, we get depressed. But let us not forget that there are also men and women all over the world who seek to address the challenges more creatively and cooperatively, who share their resources, who engage their intelligence and inventiveness to rid humanity of the ailments that afflict it.

We have reason to feel optimistic that with the knowledge that comes from science and the goodwill that must come from religions, we human beings will be able to deal with the water crisis that is staring at us. For this we must act rather than dream. There is a story about three youths who were stranded on an island in the sea where they found but a single bottle of fresh water. The question arose as to who would get to drink it. They decided they would go to sleep, and whoever had the best dream would be the privileged one. Then in the morning, the first one said he had dreamt he had won a million dollars in a lottery. The second said he had been to heaven where he experienced all the joys and riches imaginable. The third said that he so worried about the water that he simply couldn't sleep, so got up and drank the bottle empty. The moral of the story is that action, not dreaming, is what will solve our problems.

I am confident that one day with the successful tapping of limitless solar power, we will be able to provide enough energy for every man, woman, and child in every nation in the world. Likewise, I am convinced that one day we will learn to desalinate the waters of the oceans inexpensively, and to harvest rain-bearing clouds that pour down wastefully over the seas. Then we would have solved the problem of water scarcity also.

While such efforts are underway, let us keep admiring the clouds that pass over us. Let us keep rejoicing in the droplets of dew that we detect in the morning on tender leaves. Let us be in rapture when we see snow falling on rooftops and trees. Let us delight in a tall glass of ice-cold water when the sun is unduly hot. Let us share the water in our homes with thirsty friends. Let us swim where we can, and savor a sauna that is within reach. Let us continue to boil water for tea and coffee, for soup and spaghetti. Let us not be extravagant in watering our lawns and cleaning our cars. Above all, let us be grateful for the summer showers as blessings from a source beyond our control.

And in that spirit let us sing together:

Homage to Water

Oh Water, source of all the life we know,
Covering most of the earth below,
In tides you roll in highs and lows,
In sap of trees, in our blood that flows,
You are in the protoplasm first,
You cool our bodies, you quench our thirst.
In creek and brook and pond and lake
The world a wondrous place you make.
As vapors you rise, and make a link,
With land through rain, that we may drink.
No rock or stone can ever break you,
But you can slowly erode them too.
'cause of your flow in many a land,
Many a culture does proudly stand.
As steam and ice you come and go,
Your molecule is H-2-O.
Water, you're a boon indeed
You, that we so direly need.
No creature on earth without you can live.
To you, Oh Water, our homage we give.

www.ingramcontent.com/pod-product-compliance
Lightning Source LLC
Chambersburg PA
CBHW021847170526
45157CB00007B/2977